Disclaimer

The publisher of this book is by no way associated with the National Institute of Standards and Technology (NIST). The NIST did not publish this book. It was published by 50 page publications under the public domain license.

50 Page Publications.

Book Title: Camera Recognition

Book Author: Michelle P. Steves; Brian C. Stanton; Mary F. Theofanos; Dana E. Chisnell; Hannah Wald

Book Abstract: The Department of Homeland Security's (DHS) United States Visitor and Immigrant Status Indicator Technology (US-VISIT) program is a biometrically-enhanced identification system primarily situated at border points of entry such as airports and seaports. In a 2004 assessment of the quality of facial images captured by US-VISIT, the National Institute of Standards and Technology (NIST) discovered a widespread problem: many subjects were 1) not directly facing the camera and 2) had a pose angle of greater than 10 degrees. The findings of NIST's subsequent follow-up studies suggest that the camera used to capture facial images of travelers should look as much as possible like a traditional camera. Knowing where to look will help the subjects being photographed orient themselves in such a way that they are frontal to the camera – thus improving picture quality. This study explored whether participants could discern image capture devices (i.e., cameras) from other types of technology, and the attributes they relied upon to make that distinction.

Citation: NIST Interagency/Internal Report (NISTIR) - 7921

Keyword: usability; biometrics; affordance; facial recognition; cameras

NISTIR 7921

Camera Recognition

Michelle Steves
Brian Stanton
Mary Theofanos
Dana Chisnell
Hannah Wald

http://dx.doi.org/10.6028/NIST.IR.7921

NISTIR 7921

Camera Recognition

Michelle Steves
Brian Stanton
Mary Theofanos
Information Access Division
Information Technology Laboratory

Dana Chisnell
Usability Works
Boston, MA

Hannah Wald
Booz Allen Hamilton
McLean, VA

http://dx.doi.org/10.6028/NIST.IR.7921

March 2013

U.S. Department of Commerce
Rebecca Blank, Acting Secretary

National Institute of Standards and Technology
Patrick D. Gallagher, Under Secretary of Commerce for Standards and Technology and Director

Table of Contents

1 INTRODUCTION 2
2 METHOD 3
    2.1 Participants 3
    2.2 Materials 5
    2.3 Equipment 5
    2.4 Procedure 6
        2.4.1 Demographic questionnaire 6
        2.4.2 Part 1: Identifying cameras 7
        2.4.3 Part 2: Categorizing devices 7
        2.4.4 Part 3: Debrief 7
3 RESULTS 8
    3.1 Responses to "Is this a camera?" question 8
        3.1.1 Overall accuracy of participant responses 8
        3.1.2 Results for image capture devices 9
        3.1.3 Results for non-image capture devices 15
    3.2 Card Sorting 20
        3.2.1 Number of sorting categories 20
        3.2.2 Sorting category labels 21
        3.2.3 Category features and attributes 22
    3.3 Debrief 25
        3.3.1 Previous experience with cameras 25
        3.3.2 How participants identified cameras 25
4 DISCUSSION 26
    4.1 Identifying Devices 26
        4.1.1 Image capture devices (cameras) 26
        4.1.2 Non-image capture devices (not cameras) 27
        4.1.3 Phones 29
        4.1.4 Uncertainty (the "Miscellaneous" class) 29
        4.1.5 Primary camera attributes 29

4.2	Issues affecting the study	30
4.2.1	Environment and circumstances	30
4.2.2	Demographics	31
4.2.3	Technology exposure	32
5	**CONCLUSION**	**32**
6	**REFERENCES**	**34**
APPENDIX A: DEMOGRAPHIC QUESTIONS		**35**
APPENDIX B: DEVICE IMAGES USED IN THE STUDY		**36**
APPENDIX C: DATA COLLECTION SHEET		**39**
APPENDIX D: CAMERA STUDY SCRIPT		**41**
APPENDIX E: WHERE ARE VISITORS TO THE US COMING FROM?		**45**
APPENDIX F: TECHNOLOGY ADOPTION BY COUNTRY		**48**

List of Figures

FIGURE 1: A TYPE OF CAMERA USED BY US-VISIT TO CAPTURE FACIAL IMAGES 3

FIGURE 2: SELF-REPORTED ETHNICITY FOR THE STUDY POPULATION 4

FIGURE 3 *IS THIS A CAMERA?* APPLICATION 6

FIGURE 4: PERCENTAGE OF CORRECT IDENTIFICATIONS PER IMAGE CAPTURE DEVICE 14

FIGURE 5: PERCENTAGE OF CORRECT IDENTIFICATIONS PER NON-IMAGE CAPTURE DEVICE 19

FIGURE 6: NUMBER OF CATEGORIES USED BY PARTICIPANTS DURING THE CARD SORT EXERCISE 20

FIGURE 7: FROM LEFT TO RIGHT, A PHOTOGRAPHY CAMERA, A WEB CAMERA, AND A SURVEILLANCE CAMERA ... 27

FIGURE 8: NON-IMAGE CAPTURE DEVICES THAT PARTICIPANTS FREQUENTLY MISTOOK FOR CAMERAS 28

List of Tables

TABLE 1: STUDY PARTICIPANTS' SELF-REPORTED COUNTRIES OF ORIGIN 4

TABLE 2: SUMMARY OF CAMERA IDENTIFICATION RESULTS – TRUE AND FALSE POSITIVES AND NEGATIVES 9

TABLE 3: IMAGE CAPTURE DEVICE IDENTIFICATION FOR INDIVIDUAL DEVICES 10

TABLE 4: NON-IMAGE CAPTURE DEVICE IDENTIFICATION FOR INDIVIDUAL DEVICES 15

TABLE 5: CARD SORT CATEGORY LABELS, NUMBER OF TIMES USED, AND BROAD CLASSES 21

TABLE 6: PARTICIPANTS' REPORTED EXPERIENCE WITH DIFFERENT TYPES OF CAMERAS 25

TABLE 7: TECHNOLOGY ADOPTION STATISTICS FOR THE US AND TOP 15 VISITING COUNTRIES 50

EXECUTIVE SUMMARY

The Department of Homeland Security's (DHS) United States Visitor and Immigrant Status Indicator Technology (US-VISIT) program is a biometrically-enhanced identification system primarily situated at border points of entry such as airports and seaports. The US-VISIT program's goal is to advance the security of the United States and worldwide travel through information sharing and biometric solutions to facilitate identity management. The biometrics currently captured at US-VISIT primary inspection are fingerprints and a facial image. For the purposes of our study, we are interested in the latter.

In a 2004 assessment of the quality of facial images captured by US-VISIT, the National Institute of Standards and Technology (NIST) discovered a widespread problem: many subjects were 1) not directly facing the camera and 2) had a pose angle of greater than 10 degrees [2]. The findings of NIST's subsequent follow-up studies suggest that the camera used to capture facial images of travelers should look as much as possible like a traditional camera [4]. Knowing where to look will help the subjects being photographed orient themselves in such a way that they are frontal to the camera – thus improving picture quality.

This study explored whether participants could discern image capture devices (i.e., cameras) from other types of technology, and the attributes they relied upon to make that distinction. In a controlled environment, we presented participants with 50 images of small (hand-held) devices and asked participants to indicate whether or not a given device was a camera. We then asked participants to group the devices into 2 to 5 categories and list the attributes they had used as a basis for assigning devices to each group.

We did not attempt to simulate the environment or context in which US-VISIT cameras are normally used – specifically, a busy, active, visually and aurally crowded port of entry. However, if individuals in a lower-stress environment were not able to identify image capture devices, it is unlikely that people in a more stressful situation will be able to do so.

Our findings indicate that the more an image capture device resembles a traditional photography camera – generally rectangular with a round, projecting lens and a flash – the easier it is for individuals to notice that device and recognize it. Once this recognition occurs, it is then expected that individuals could more confidently position themselves properly to have their pictures taken.

1 INTRODUCTION

The Department of Homeland Security's (DHS) United States Visitor and Immigrant Status Indicator Technology (US-VISIT) program is a biometrically-enhanced identification system primarily situated at border points of entry such as airports and seaports. The US-VISIT program's goal is to advance the security of the United States and worldwide travel through information sharing and biometric solutions to facilitate identity management. The biometrics currently captured at US-VISIT primary inspection are fingerprints and a facial image. The fingerprint component of the system uses automated matching along with manual match verification. The face image capture process does not include automated face recognition but relies on human verifiable traveler history.

A face image quality assessment of airport ports of entry performed in 2004 found a number of quality issues in approximately 1.5 million facial images captured by US-VISIT [2]. One issue was that only 5% of the time were the subjects being photographed directly facing the camera, that is, they were frontal to it, and approximately 70% of facial images had a pose angle of greater than 10 degrees.

As the result of this assessment US-VISIT has embarked on a program for face image quality improvement. One aspect of this effort is the identification of usability and human factors issues that may impact face image capture. The National Institute of Standards and Technology's (NIST) usability and biometrics team was asked to identify any usability and human factors considerations that may improve the capture of face images at the airports.

This study, which is part of that effort, focuses specifically on the camera used to take pictures of travelers at ports of entry. In a previous study, we recommended that the camera used to capture facial images of travelers should look as much as possible like a traditional camera [4]. We suspect that the device currently used by US-VISIT – a webcam on the end of a flexible gooseneck, as shown in Figure 1 – may not be recognizable to some travelers as a camera, which makes them unsure of where to look when having their picture taken. Knowing where to look will help the subjects being photographed orient themselves in such a way that they are frontal to the camera – thus improving picture quality.

Figure 1: A type of camera used by US-VISIT to capture facial images

In this between participants study[1], we explored how well participants could recognize cameras from an array of devices. We also attempted to identify what features make a particular device recognizable as a camera. In essence, our study was designed to answer two questions:

1. In a visually and aurally quiet environment (i.e., one with no distractions), can participants distinguish camera devices from other pieces of technology?

2. What are the attributes of image capture devices that make them recognizable as cameras?

By answering these questions, we hope to provide recommendations that will inform the design and/or selection of cameras to be used by the US-VISIT program for capturing facial images at ports of entry. We also anticipate that an easily recognizable camera will be an important part of any "self-service" facial image capture solution that US-VISIT may implement in the future.

2 METHOD

2.1 Participants

Researchers solicited for volunteer participants for this study in the DC area. There were a total of 86 participants in the camera recognition study. 42 were male and 44 were female. The average age of the participants was 39 years, with the youngest participant being 20 years old, and the oldest being 69 years old. Participants also reported their ethnicity on the demographic survey; Figure 2 displays the demographic breakdown of the study population.

---

[1] Between subjects design is a study design where every participant is subjected to a single treatment. It is also referred to as an independent measures design.

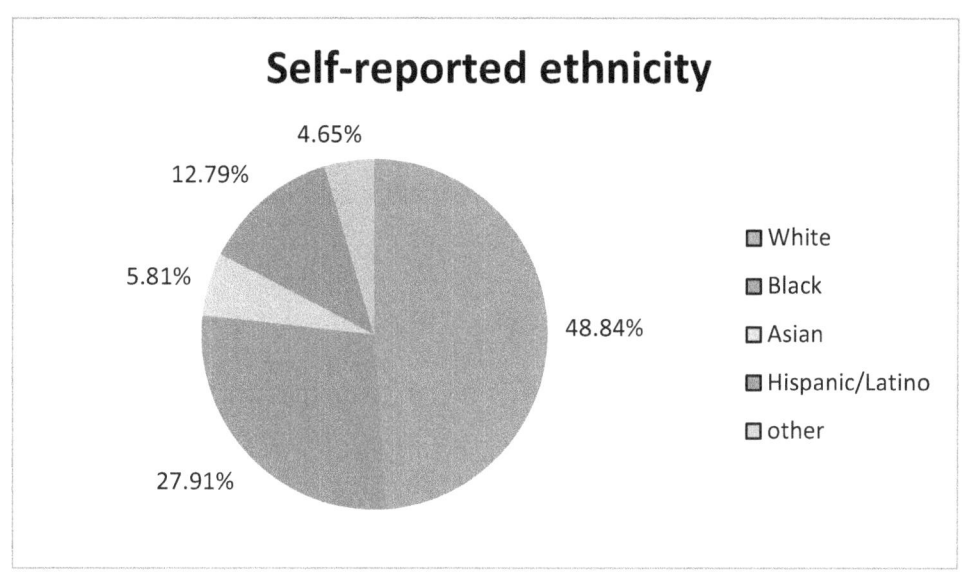

Figure 2: Self-reported ethnicity for the study population

Our participants also reported their countries of origin. As shown in Table 1, the vast majority were American (76 of 86, or 88.37%), but there were a few participants from other countries as well.

Table 1: Study participants' self-reported countries of origin

Country	Count	%
USA	76	88.37
Vietnam	2	2.33
Canada	1	1.16
China	1	1.16
Columbia	1	1.16
El Salvador	1	1.16
Gabon	1	1.16
Haiti	1	1.16
Kyrgyzstan	1	1.16
Republic of the Congo	1	1.16

Finally, participants were asked if they had had their biometrics captured before and whether that included facial image capture. Fifty-two (or 60.47%) said they had biometrics captured previously and 8 (or 9.30%) reported that this biometrics data included facial image data.

2.2 Materials

The following materials were used in this study:

- Informed consent form explaining the study to participants and how the data would be used

- Demographic questionnaire asking participants about their age, ethnicity, occupation, and previous experience with biometrics data capture, contained in Appendix A: Demographic Questions

- Script for the researcher that stipulated what the researcher should say, directions for moving from station to station in the lab space, and pointers for data collection, contained in Appendix D: Camera Study Script

- Data collection sheets on which the researcher tracked participants' behaviors and comments, and recorded observations, contained in Appendix C: Data Collection Sheet

2.3 Equipment

During the study, participants used a personal computer with an application that presented them with images of 50 devices[2], one at a time. Appendix B: Device Images used in the Study contains the images. For each image, participants were prompted to answer the question "Is this a camera?" Participants could respond "Yes" or "No" by clicking the appropriate checkbox. A sample screenshot from this application can be seen in Figure 3.

---

[2] The use of certain commercial equipment in this study is not intended to imply recommendation or endorsement by National Institute of Standards and Technology, nor is it intended to imply that the equipment shown is necessarily the best available for the purpose.

Figure 3 Is this a Camera? application

Twenty-nine of the 50 devices (or 58%) were actually cameras. The application would record participants' positive ("Yes") or negative ("No") responses to each image, as well as whether participants were correct in identifying the device shown as a camera (or not).

In addition, participants were issued 50 laminated cards measuring 5.1 cm by 7.6 cm (2 in. by 3 in.), each bearing one of the images previously used in the "Is this a camera?" application. Participants sorted these cards into categories and labeled the categories using sticky notes (Section 2.4.3 contains further details).

2.4 Procedure

2.4.1 Demographic questionnaire

Each participant signed an informed consent form and filled out a demographic questionnaire (Appendix A: Demographic Questions) before proceeding with the study as described in the following subsections. The demographic questionnaire asked participants for their age, gender, ethnicity, and profession. It also asked participants if they had previous experience with biometric data capture, and if so, what type of data was captured, e.g., fingerprints, facial image, and so on.

2.4.2 Part 1: Identifying cameras

Participants used a computer-based application to view 50 images of devices. Each was accompanied by the question "Is this a camera?" The participants would answer "Yes" or "No" by clicking in the corresponding check box on the screen. The application recorded their choices. It also recorded whether or not a participant correctly identified a device as a camera or something else, i.e., whether they entered "Yes" for a camera device or "No" for a non-camera device.

2.4.3 Part 2: Categorizing devices

After participants completed the computer-based portion of the study, they were given 50 cards, each measuring 5.1 cm by 7.6 cm (2 in. by 3 in.) and each bearing the image of one of the devices they had seen on the computer. The researcher instructed participants to sort the cards into 2 to 5 categories of their choosing. Participants were given five minutes for this task, but could take more time if they wished.

Once the participants finished sorting their cards, the researcher gave them a sticky note pad and pen and asked them to make a label for each category. Then the researcher asked participants to list the features or attributes of each group. The researcher recorded the participants' categories and associated attributes on the data sheet.

2.4.4 Part 3: Debrief

After participants completed parts 1 and 2, the researcher debriefed them about their previous experience with cameras (e.g., whether or not they had ever operated various types of cameras). In addition, the researcher asked participants about their experiences with biometric data capture (as in, being the subject of biometric data capture, not the operator of the biometric device) and whether that included a facial image capture. The researcher used the checklist at the bottom of the data sheet to record participant responses. The researcher then inquired how the participant identified a particular device as a camera. Finally, the researcher thanked the participants for taking part in the study.

3 RESULTS

Researchers collected the following data from participants:

- Results of "Is this a camera?" application – whether the participant answered positively or negatively, and whether they correctly identified a given device as a camera (or not).

- Card sorting category labels and features/attributes

- Participants' previous experience with cameras

- How participants identified a particular device as a camera

The information collected is described in detail in the following subsections.

3.1 Responses to "Is this a camera?" question

When participants used the computer application that presented them with images of 50 devices, they classified each device as a camera or non-camera (which we refer to as "image capture" or "non-image capture" devices). This section describes the results of that exercise.

3.1.1 Overall accuracy of participant responses

Table 2 is a matrix showing participants' overall accuracy when identifying image capture or non-image capture devices while using the "Is this a camera?" application. The top row shows the percentage of participant responses that were "True Positive" or "True Negative," meaning that they were accurate in regards to whether the device presented was a camera or not. The bottom row shows the percentage of false responses, or cases in which participants incorrectly identified a non-image capture device as an image capture device (or vice-versa).

Table 2: Summary of camera identification results – true and false positives and negatives

	Positives ("Yes")	Negatives ("No")
True (participant response was correct)	88.69%	68.05%
False (participant response was incorrect)	11.31%	31.95%

The data show that participants were fairly accurate when positively identifying devices that functioned as cameras – nearly 89% of their positive responses to the "Is this a camera?" question were correct. Participants only gave false positive responses (in which they identified non-image capture devices as image capture devices) approximately 11% of the time. However, they were somewhat less accurate when identifying non-image capture devices; when presented with these devices by the computer application, they only identified non-image capture devices as such (through a true negative response) approximately 68% of the time, and mistook the non-image capture devices for image capture devices almost 32% percent of the time. That is, 32% of the time that subjects were presented with images of actual cameras, they did not recognize them as such.

3.1.2 Results for image capture devices

Table 3 provides the count and corresponding percentage of correct and incorrect identification responses for each image capture device shown in the study (on the "Is this a camera?" application and later on the 5.1 cm by 7.6 cm cards). Of the 50 devices presented to participants, 29 were actually image capture devices (i.e., cameras).

Note that some devices received the same number of correct identification responses – for example, the first eleven devices shown in Table 3 were correctly identified 98.84% of the time. No additional ranking within a particular response count (i.e., among those eleven devices) is intended by the tabular and graphical representations of the data – all are equivalent. Figure 4, following the table, provides a visual reference for the percentage of correct identifications (that is, "Yes" responses to the "Is this a camera?" question) for each device depicted in the table.

Table 3: Image capture device identification for individual devices

Device Reference Number	Device Image	Correct Responses		Incorrect Responses	
		#	%	#	%
Ref 1		85	98.84%	1	1.16%
Ref 2		85	98.84%	1	1.16%
Ref 3		85	98.84%	1	1.16%
Ref 4		85	98.84%	1	1.16%
Ref 5		85	98.84%	1	1.16%
Ref 6		85	98.84%	1	1.16%
Ref 7		85	98.84%	1	1.16%
Ref 8		85	98.84%	1	1.16%

Device Reference Number	Device Image	Correct Responses #	Correct Responses %	Incorrect Responses #	Incorrect Responses %
Ref 9		85	98.84%	1	1.16%
Ref 10		85	98.84%	1	1.16%
Ref 11		85	98.84%	1	1.16%
Ref 12		84	97.67%	2	2.33%
Ref 13		84	97.67%	2	2.33%
Ref 14		83	96.51%	3	3.49%
Ref 15		83	96.51%	3	3.49%

Device Reference Number	Device Image	Correct Responses #	Correct Responses %	Incorrect Responses #	Incorrect Responses %
Ref 16		83	96.51%	3	3.49%
Ref 17		82	95.35%	4	4.65%
Ref 18		82	95.35%	4	4.65%
Ref 19		82	95.35%	4	4.65%
Ref 20		82	95.35%	4	4.65%
Ref 21		80	93.02%	6	6.98%
Ref 22		72	83.72%	14	16.28%

Device Reference Number	Device Image	Correct Responses		Incorrect Responses	
		#	%	#	%
Ref 23		69	80.23%	17	19.77%
Ref 24		63	73.26%	23	26.74%
Ref 25		63	73.26%	23	26.74%
Ref 26		61	70.93%	25	29.07%
Ref 27		52	60.47%	34	39.53%
Ref 28		39	45.35%	47	54.65%
Ref 29		33	38.37%	53	61.63%

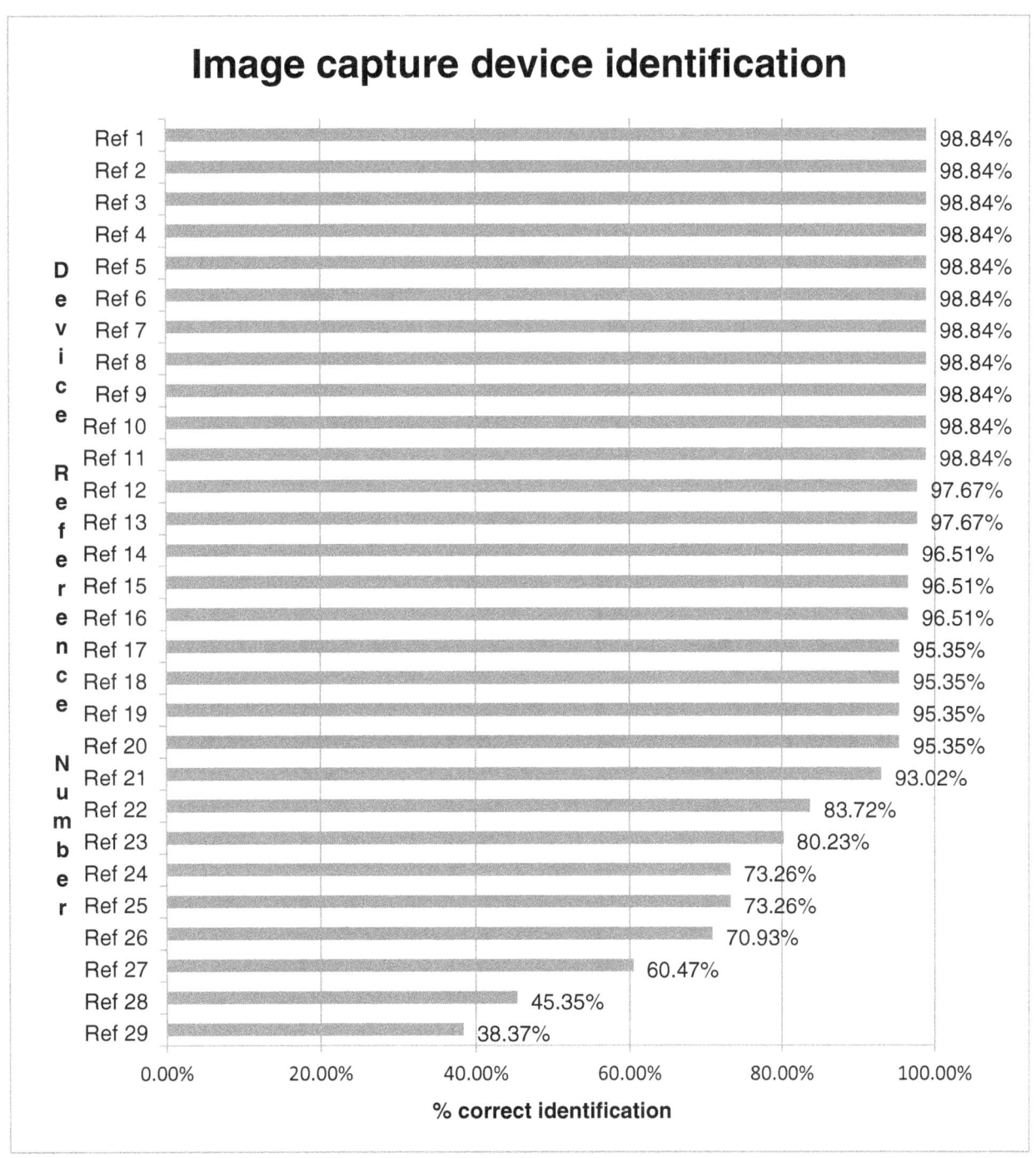

Figure 4: Percentage of correct identifications per image capture device

As mentioned in Section 3.1.1, participants were generally highly successful in positively identifying image capture devices. Participants had a success rate of over 90 percent when identifying 21 of the 29 devices: for 11 of those devices, the success rate was over 98%. For only 2 devices did participants have a success rate of less than 50%.

3.1.3 Results for non-image capture devices

Table 4 and Figure 5 below are similar to those in Section 3.1.2, except that they deal with *non*-image capture devices – of which there were 21 in the set of 50 presented to participants – and count correct *negative* (rather than positive) responses to the "Is this a camera?" question.

Table 4: Non-image capture device identification for individual devices

Device Reference Number	Device Image	Correct Responses		Incorrect Responses	
		#	%	#	%
Ref 30		85	98.84%	1	1.16%
Ref 31		82	95.35%	4	4.65%
Ref 32		81	94.19%	5	5.81%
Ref 33		79	91.86%	7	8.14%

Device Reference Number	Device Image	Correct Responses		Incorrect Responses	
		#	%	#	%
Ref 34		77	89.53%	9	10.47%
Ref 35		68	79.07%	18	20.93%
Ref 36		67	77.91%	19	22.09%
Ref 37		67	77.91%	19	22.09%
Ref 38		64	74.42%	22	25.58%
Ref 39		63	73.26%	23	26.74%
Ref 40		62	72.09%	24	27.91%
Ref 41		61	70.93%	25	29.07%

Device Reference Number	Device Image	Correct Responses		Incorrect Responses	
		#	%	#	%
Ref 42		59	68.60%	27	31.40%
Ref 43		53	61.63%	33	38.37%
Ref 44		45	52.33%	41	47.67%
Ref 45		44	51.16%	42	48.84%
Ref 46		40	46.51%	46	53.49%
Ref 47		40	46.51%	46	53.49%

Device Reference Number	Device Image	Correct Responses		Incorrect Responses	
		#	%	#	%
Ref 48		39	45.35%	47	54.65%
Ref 49		32	37.21%	54	62.79%
Ref 50		21	24.42%	65	75.58%

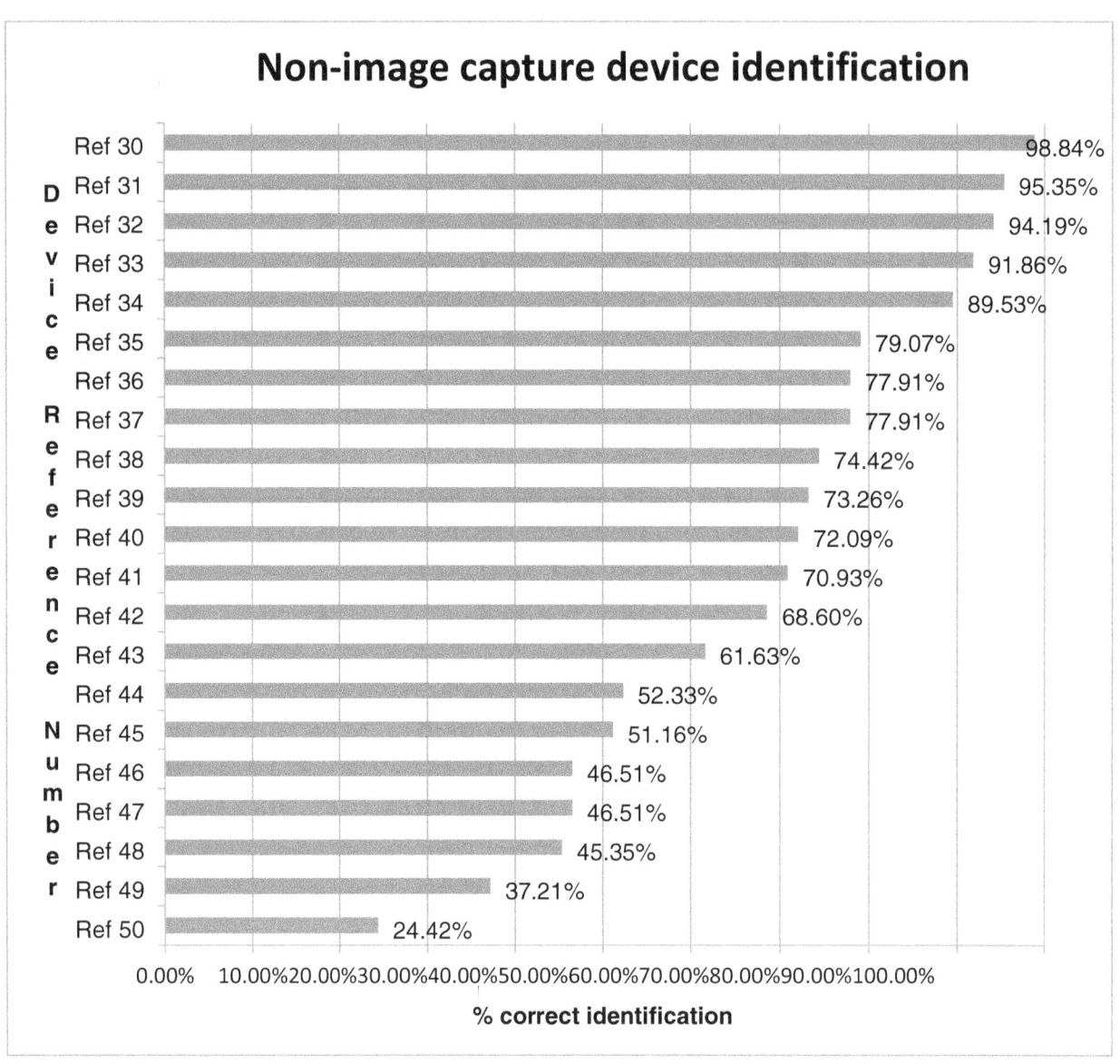

Figure 5: Percentage of correct identifications per non-image capture device

As seen in the chart above, participants had a success rate of greater than 90 percent when identifying 4 of the 21 non-image capture devices as such. For 7 devices in the non-image capture group, participants had a success rate of more than 70% but less than 90%. However, they had a success rate of only slightly over 50% with two devices, and less than 50% with 5 of them. For one device, the success rate was 24.42% – the lowest of any device shown to participants.

3.2 Card Sorting

The following subsections present data from the card-sorting exercise that participants performed after using the "Is this a camera?" application. The data include the number of categories into which participants sorted cards, how participants labeled those categories, and the attributes/characteristics participants used to distinguish between categories.

3.2.1 Number of sorting categories

Participants were asked to sort their cards into 2 to 5 categories. Figure 6 shows the distribution of the number of categories into which participants sorted their cards. As shown, many participants preferred to separate rather than consolidate: on average, participants used 4.3 categories, with the majority of participants (65.12%) using five categories.

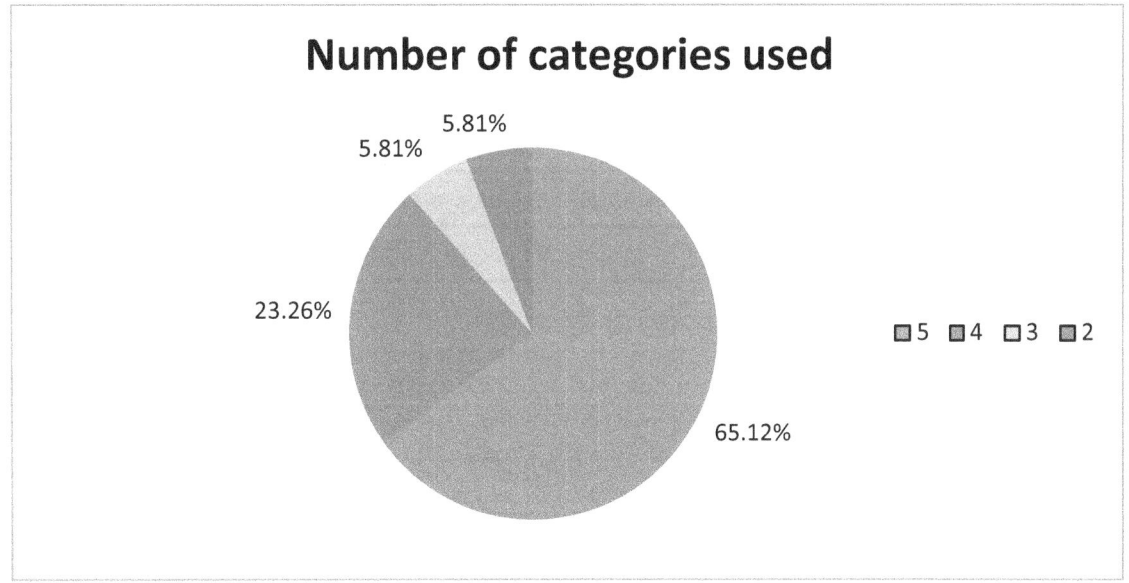

Figure 6: Number of categories used by participants during the card sort exercise

3.2.2 Sorting category labels

Participants were asked to label the categories they made using sticky notes. Participants created a total of 31 labels. In an effort to consolidate the data, we sorted participant labels into four broad classes, based on participant category labels. These four classes are: Cameras, Phones, Non-Cameras, and Miscellaneous. Categories in the "Miscellaneous" class have ambivalent or nonspecific labels (e.g., "Unknown/no clue," "Miscellaneous," "Camera-like device"): participants essentially used these categories for devices they could not place in their other categories, or devices they could not identify. Table 5 below displays the labels used, the number of times they were used, and the broad classes into which we sorted them.

Table 5: Card sort category labels, number of times used, and broad classes

Broad Classes	Category Labels	Number of Times used
Cameras	Camera/traditional camera	76
	Web/computer camera	52
	Monitoring/surveillance device	12
	Image capture device (all)	8
	Looks like a (it is a) camera - not traditional	2
	Portable webcam	2
	Non-traditional photography device	2
	Watch camera	2
	Robotic camera	1
	Low quality image capture	1
	Camcorder	2
	Specialized / industrial (such as DMV or other ID)	1
Phones	Cell phone/smart phone	67
	Landline phone	1
Non-Cameras	Portable digital media player/MP3	43

Broad Classes	Category Labels	Number of Times used
	No camera	14
	Other electronic devices (including flash drives)	13
	Digital storage	6
	Radio	2
	Scanner	2
	No Lens	1
	Bluetooth	1
	Glucose meter	1
	Camera accessory	1
Miscellaneous	Unknown/no clue	38
	Miscellaneous	15
	Could be a camera - possible lens	3
	Handheld	4
	Appearance did not suggest camera - may have had a lens/Not sure if it was a camera	2
	Maybe a camera - couldn't find the lens	2
	Camera-like device	1
	Form over function	1
	Recreational	1
TOTAL	31 Labels	380

3.2.3 Category features and attributes

This section lists the attributes participants most commonly used to categorize devices. We have divided the attributes into the broad classes used in Section 3.2.2.

3.2.3.1 Cameras

When asked during the debriefing, 67 of the 86 participants (or 77.9%) reported that they associated the "lens" attribute with the categories in the Cameras class (see Section 3.3.2). "Lens" was, in fact, the attribute that participants most frequently associated with Camera categories. The other most commonly used attributes were:

- Shape[3]

- Buttons/lack of buttons

- Branding

- Has a base/stand/legs/swivels (specifically for Web or surveillance cameras)

- Flash

Participants reported using other attributes as well, but with much less frequency than the ones listed above, reflecting the fairly broad set of participant-labeled categories for image-capture devices.

3.2.3.2 Phones

Participants provided a relatively short list of defining attributes for devices they categorized as phones. The four most common attributes associated with categories in this class are:

- Buttons/keypad

- Screen

- Shape

- Lens

Participants used the "lens" attribute for devices in the Phone class because – as many of them observed – most newer cell phones have cameras in them. Some participants reported

---

[3] Participants also commonly used the "shape" attribute for categories in the Phones and Non-Image Capture Device class: participants used it in the sense of "the shape of this device makes me think it is _____."

that they assumed cell phones contain a camera and classified them as image capture devices without looking for a lens feature.

3.2.3.3 Non-Cameras

Participants associated many different attributes with devices in the Non-Cameras class. The attributes they reported using most frequently were:

- No lens
- USB connector
- Buttons suggesting functionality other than a camera (e.g., those for an MP3 player: play, rewind, fast forward)
- Shape (suggesting some functionality other than a camera)
- Screen or display (or lack thereof) suggesting some functionality other than a camera

3.2.3.4 Miscellaneous

Participants' attributes for devices in the Miscellaneous class typically reflected their ambivalence regarding the functionality of the devices, or whether or not the devices might have a lens that they were unable to see. Participants frequently reported associating the following attributes with devices in this class:

- Don't recognize/not sure
- No idea (what it is)
- Could be some something else entirely (than a camera device)
- No lens
- May have a lens (but couldn't find it)

3.3 Debrief

3.3.1 Previous experience with cameras

After participants completed the device identification and grouping exercises, the researcher asked them about their previous experience using various types of cameras. Participant responses are summarized in Table 6.

Table 6: Participants' reported experience with different types of cameras

Usage Experience	Number of participants reporting experience
"Point and Shoot" type camera	82
Phone camera	69
Video camera	56
Webcam	40
Single Lens Reflex (SLR)	29
No camera use	0

All 86 participants had experience with one or more types of cameras: 82 (95.35%) of participants had used a "traditional" point-and-shoot camera, 69 (80.23%) had used a cell phone camera, 56 (65.12%) had used a video camera, and 40 (46.51%) of them had used a webcam.

3.3.2 How participants identified cameras

During the debriefing, the researcher asked participants what criteria they used to determine whether a device was a camera or not. 67 of 86 participants (77.9%) reported that they used the presence or absence of a lens as an indicator of whether a particular device could capture an image.

4 DISCUSSION

The goal of our study was to determine how people distinguish image capture devices from other kinds of devices. This section deals with our findings on how people distinguish (or fail to distinguish) between cameras and other devices. We also address issues affecting the study, such as differences between our study population and the foreign nationals who undergo the US-VISIT biometric collection process at US ports of entry.

4.1 Identifying Devices

4.1.1 Image capture devices (cameras)

When attempting to determine whether or not a particular device was a camera, nearly 78% of participants (as described in Section 3.3.2) reported that they looked for a lens. This assertion is borne out by the results of the card sorting exercise: the attribute most commonly associated with categories in the Cameras class (which had labels such as "Camera/traditional camera," "Web/computer camera," and "Monitoring/surveillance device") was a lens (see Section 3.2.2).

Other attributes participants frequently associated with cameras were shape (in the sense of "the shape of this device makes me think it is a camera"); buttons; branding (as in "it has the name of a camera company on it"); the inclusion of a base or stand and/or the apparent capacity to swivel; and a flash.

By and large, the participants considered a lens to be the critical, defining element of a camera, and their high overall success rate when attempting to positively identify cameras – almost 89%, as noted in Section 3.1.1 – indicates that the "look for a lens" strategy was a valid one. However, it is important to note that participants drew distinctions between different types of cameras based on the secondary attributes listed above: these distinctions are reflected in the three most common category labels participants created during the card sorting exercises. These labels were "Camera/traditional camera" (76 times), "Web/computer camera" (52 times), and "Monitoring/surveillance device" (12 times).

While all the devices in these categories obviously had common attributes participants associated with cameras (most likely the lens, since many participants regarded a lens as a defining attribute for an image capture device), they also had *different* attributes that participants seemed to consider noteworthy. One likely example is the shape of the camera: photography cameras, whether digital or traditional, have a very different profile from most web cameras, and both differ in shape from surveillance cameras (as shown in Figure 7). The three types of cameras have different functions and tend to be used in different contexts.

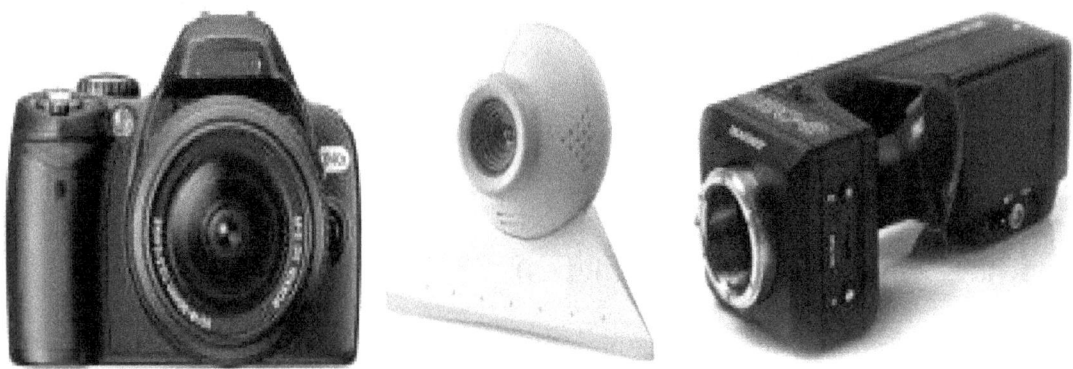

Figure 7: From left to right, a photography camera, a web camera, and a surveillance camera

Finally, while 76 of the 86 study participants (or 88.37%) created a card sorting category for photography cameras ("Camera/traditional camera"), only 52 (60.47%) created a category for webcams ("Web/computer camera"). This indicates that when participants thought about cameras, the first thing that came to mind for most of them was a photography camera. It also seems to indicate that photography cameras were more easily recognizable to a larger number of participants than were web cameras.

4.1.2 Non-image capture devices (not cameras)

Participants created a variety of categories with associated attributes for items in the Non-Cameras class, but all of these devices had one attribute in common: they lacked a lens. In fact, "no lens" was the attribute participants most frequently associated with device categories in this class. This is understandable, given that most participants appeared to think of a lens as the primary, defining attribute of a camera.

As with camera devices, participants considered other attributes when trying to identify or confirm the nature of the device they were examining. A number of participants considered certain kinds of screens/displays as an indicator that a particular device was a camera or not. If the device otherwise resembled a digital photography camera, some participants considered the *lack* of a screen or display to mean that it was something else. Finally, participants frequently took the inclusion of a USB connector as an indication that a device was something other than a camera.

What is particularly interesting is that participants often noted two attributes that they also used to identify cameras: the shape of the device and any buttons on it. However, they believed these features suggested that a device was something *other* than a camera. For example, if the device was square and had "play," "rewind," and "fast-forward" buttons but no "record" button, it probably wasn't a camera.

This indicates that while the presence or absence of a lens was – for our participants – the primary determinant of whether a device was or was not a camera, the shape of the device and the buttons on it were also important indicators.

However, participants were not as successful at correctly identifying non-cameras as they were at identifying cameras: they had a success rate of only 68.05%, as opposed to a success rate of 88.69% when positively identifying cameras. In fact they mistakenly identified non-image capture devices as cameras 31.95% of the time. Device Ref 49 and Ref 50, shown in Figure 8 below, are examples of non-image capture devices that participants often mistook for cameras (62.79% and 75.58% of the time, respectively). This is because they each had a circular feature that could easily be mistaken for a lens and a shape typical of a point-and-shoot camera device.

Figure 8: Non-image capture devices that participants frequently mistook for cameras

4.1.3 Phones

Participants typically placed cellular phone devices into a category unto themselves, labeled either "Cell phone" or "Smart phone." Most participants associated some combination of a few distinct attributes with cell phones: most often they noted the buttons or number keypad, some kind of screen or display, the shape of the device, and a lens. Many participants made an observation along the lines of "most cell phones now have cameras in them."

As previously mentioned, most participants placed phones in a category by themselves, even if they appeared to have (or were assumed to have) image capture capabilities. Some participants, however, placed phones (or rather, devices they considered to be phones) without a discernible lens into a category all by themselves. A few other participants, assuming that *all* mobile phones have image capture capabilities, placed them in the same category with cameras. This "lumping" of phones with cameras was more the exception than the rule.

4.1.4 Uncertainty (the "Miscellaneous" class)

The final class of devices participants categorized during the card sorting exercise was the Miscellaneous class. We placed in this class device categories that were vague or reflected participants' ambiguity about the devices (e.g., "Unknown/no clue," "Miscellaneous").

Since participants were trying to distinguish between cameras and other devices, the first thing many of them did was look for a lens. The devices participants put into Miscellaneous categories often had no lens or – in some cases – had features that participants thought were "maybe" or "possibly" a lens. This is reflected in some of the category labels: "Could be a camera – possible lens," "Appearance did not suggest camera – may have had a lens," and "Maybe a camera – couldn't find the lens."

Participants also tried to identify the function of these devices based on their shape and any visible buttons, but found these features so unfamiliar or confusing that they could not even attempt an educated guess.

4.1.5 Primary camera attributes

Our findings strongly indicate that when our participants were presented with visually uncluttered images of various devices in a quiet, non-stressful environment, they were very good at positively identifying cameras. Overall, they would look for a discernible lens on the device they were examining: if it had no lens, it could not be a camera. Even if a device's other attributes (such as its shape and visible buttons) suggested that it might be a camera, participants were extremely reluctant to identify it as such without a lens.

Although the lens was the defining attribute of a camera, participants generally looked at the shapes of and visible buttons on devices as well when examining them. They used the specific nature and details of these features to try and identify the specific function and purpose of the devices they examined. This is an important consideration, because if a device appears to have a lens, its shape and buttons can serve to confirm whether it is a camera or something else (e.g., a cell phone).

4.2 Issues affecting the study

This study is part of a larger effort to develop recommendations and best practices the Department of Homeland Security can implement to make the US-VISIT biometric collection process more usable for both foreign nationals who undergo it and the customs officers who oversee it. Our study could not perfectly replicate the user population (foreign nationals), environment (customs at US ports of entry), or circumstances of the actual US-VISIT process. We address the dissimilarities between the study and real-world operational environment in the following subsections.

4.2.1 Environment and circumstances

Participants in this study performed their assigned tasks in a relatively quiet, low-stress environment. The actual environments in which US-VISIT cameras are deployed are considerably noisier and more cluttered, with a larger number of distractions to contend with than in a controlled laboratory environment. Foreign nationals undergoing the US-VISIT process may also suffer from travel fatigue, time pressure, and other stresses that affect their performance during the US-VISIT process – including how they perform when identifying the camera that will capture their facial images.

In addition, our study population may have been primed for their task by certain words used in the instruction sets, an influence that is not present in the experience of foreign nationals in real operational environments.

Finally, if our study participants provide any reliable indication, foreign nationals may perceive having their picture taken for US-VISIT differently than they do having their picture taken for other identification purposes. While more than half of our study participants (52, or 64.47%) said that they had had their biometric data captured in the past, only 8 (9.30%) reported having had their facial biometrics captured. This is a somewhat unexpected result, as most adults in the United State possess some form of picture identification that is kept on record with a government agency or place of employment (e.g., driver's license, passport, employer facility ID). This discrepancy may be due to a semantic issue: it is possible that participants did not perceive their identification pictures as "biometrics."

4.2.2 Demographics

DHS collects and publishes demographic data on visiting foreign nationals. We use their data from 2009 (see Appendix E: Where are Visitors to the US Coming From?) as a basis for comparison between our study population and the population of individuals who undergo the US-VISIT process at ports of entry. While both populations are composed of almost equal numbers of males and females, they are dissimilar in other respects, such as age and (especially) individuals' countries of origin.

According to DHS, 60% of all non-resident admissions in 2009 were of people aged 25 through 54: another 21% of admissions were of people aged 55 or over. In our study population, 82.5% of participants were between 20 and 54 years of age (inclusive), and only 17.5% of our participants were 55 or older. Our population was, on average, younger than that of foreign visitors to the US.

As one might suspect, our study participants' countries of origin (shown in Table 1) did not reflect that of visitors to the US – most of our participants were American, while foreign visitors, by definition, are not. DHS statistics indicate that "top visiting countries" in 2009 – those from whom the greatest number of foreign visitors to the US originated – were the Mexico, United Kingdom, Japan, Germany, France, Italy, Brazil, Australia, the Netherlands, and Spain.

4.2.3 Technology exposure

By combining the aforementioned DHS data with technology adoption data from the World Bank (see Appendix F: Technology Adoption by Country), we can compare visiting foreign nationals' average level of technology exposure with that of the study population.

Most people from the "top visiting countries" (listed in Section 4.2.2) are on par with or better than individuals in the United States in terms of their access and exposure to technology (e.g., personal computers, the Internet, or telecom technology). However, residents of most other countries around the world have less access to personal computers and the Internet than a typical individual in the US does. Mobile phones are an exception: they have achieved a higher rate of technological penetration worldwide than personal computers or other telecommunications technologies (see the table in Appendix F: Technology Adoption by Country). We can gather from this that the vast majority of foreign visitors to the US would be able to recognize a mobile phone as easily as our study participants did, but may not be (on average) as familiar with web cameras.

5 CONCLUSION

This study explored whether participants could discern image capture devices from other types of technology and the attributes they relied upon to make that distinction. In a controlled environment, we presented participants with 50 images of small (hand-held) devices and asked participants to indicate whether or not a given device was a camera. We then asked participants to group the devices into 2 to 5 categories and list the attributes they had used as a basis for assigning devices to each group.

We did not attempt to simulate the environment or context in which US-VISIT cameras are normally used – specifically, a busy port of entry, many stations with various pieces of equipment, a list of instructions for the exit procedure, and the time pressures and other stresses of travel. However, if individuals in a lower-stress environment were not able to identify image capture devices, it is unlikely that people in a more stressful situation will be able to do so.

Participants in our study were highly successful at correctly identifying image capture devices, but were somewhat less successful at correctly identifying non-image capture devices. Most of the participants indicated that, when trying to determine whether a particular device was a camera, the first thing they would do was look for a lens. This assertion was borne out in the device categorizing exercise: participants tended to categorize devices as cameras if they had a discernible lens, and as something else if they did not have a lens. An exception was cell phones, which – as many participants observed – have cameras in them. Some participants put cell phones in the same category with cameras, but most put them in a separate category.

Participants also considered a device's shape and visible buttons. They used these attributes to try and draw inferences about the specific function of the device, whether they believed it was a camera or not. Unfamiliar shapes or buttons seemed to confuse participants to the point where they could not determine a device's functionality.

This leads us to conclude that most individuals can easily recognize an image capture device if it has a prominent lens, but other features of the device are important as well. People will look closely at the shape of and controls on a device in order to determine its function. Also, the "secondary" attributes of a camera (such as its shape, buttons, and a flash) may make it especially distinctive, which is important in a high-stress environment where many distractions are present. A US-VISIT camera that strongly resembles a traditional photography camera – generally rectangular with a round, projecting lens and a flash – should be not only easily recognizable to most people, but unmistakable.

6 REFERENCES

[1] EconStats, (2012). *Personal computers (per 100 people)*. Data retrieved April 25, 2012 from EconStats database. Retrieved from website: http://www.econstats.com/wdi/wdiv_597.htm

[2] Monger, R. and Barr, M. (2010, April) Nonimmigrant Admissions to the United States: 2009, Annual Flow Report, Office of Immigration Statistics. Retrieved from website: http://www.dhs.gov/xlibrary/assets/statistics/publications/ni_fr_2009.pdf

[3] Nadel, L. (2007, November). *Approaches to Face Image Capture at US-VISIT Ports of Entry*. Paper presented at NIST biometric quality workshop II, Gaithersburg, MD. Retrieved from http://biometrics.nist.gov/cs_links/quality/workshopII/proc/nadel_Approaches_to_Face_Image_Capture_at_US-VISIT_POEs.pdf

[4] Theofanos, M., Stanton, B., Sheppard, C., Micheals, R., Libert, J., & Orandi, S. US Department of Commerce, Technology Administration, National Institute of Standards and Technology. (2008). *Assessing face acquisition* (NISTIR 7540). Retrieved from National Institute of Standards and Technology website: http://zing.ncsl.nist.gov/bioUSa/docs/face_IR-7540.pdf

[5] US Department of Homeland Security, (2009). *Yearbook of immigration statistics: 2009, supplemental table 1*. Retrieved from website: http://www.dhs.gov/xlibrary/assets/statistics/yearbook/2009/nimsuptable1d.xls

[6] World Bank. (2012). Data retrieved April 20, 2012 from World Development Indicators Online (WDI) database. Retrieved from website: http://data.worldbank.org/data-catalog/world-development-indicators

APPENDIX A: DEMOGRAPHIC QUESTIONS

Demographic Questions

1. Age: _____

2. Gender: (circle one)   male   female

3. Ethnicity: _____

4. Profession: _____

5. Have you ever had your biometrics captured before? (circle one)   yes   no

If yes check all that apply:

___ Fingerprinted with ink/paper

___ Fingerprinted electronically

___ Palm Print

___ Eye Scan

___ Face Image

___ Voice

___ Hand geometry

APPENDIX B: DEVICE IMAGES USED IN THE STUDY

Image Capture Devices				
Ref # 1	Ref # 2	Ref # 3	Ref # 4	Ref # 5
Ref # 6	Ref # 7	Ref # 8	Ref # 9	Ref # 10
Ref # 11	Ref # 12	Ref # 13	Ref # 14	Ref # 15
Ref # 16	Ref # 17	Ref # 18	Ref # 19	Ref # 20
Ref # 21	Ref # 22	Ref # 23	Ref # 24	Ref # 25

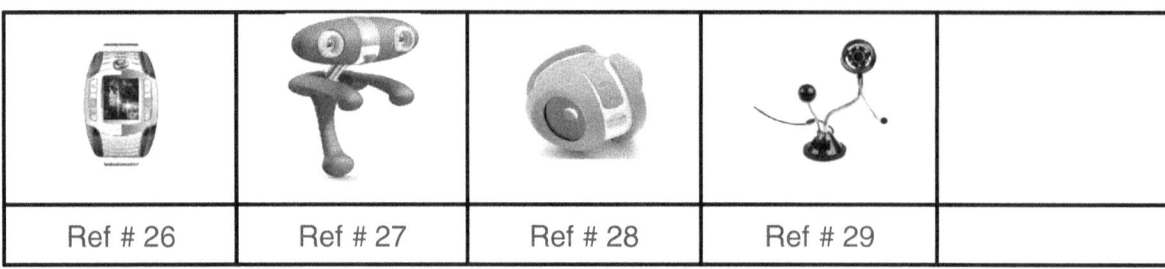

Ref # 26	Ref # 27	Ref # 28	Ref # 29	

Non-Image Capture Devices				
Ref # 30	Ref # 31	Ref # 32	Ref # 33	Ref # 34
Ref # 35	Ref # 36	Ref # 37	Ref # 38	Ref # 39
Ref # 40	Ref # 41	Ref # 42	Ref # 43	Ref # 44
Ref # 45	Ref # 46	Ref # 47	Ref # 48	Ref # 49

Ref # 50				

APPENDIX C: DATA COLLECTION SHEET

☐ **Cameras identified** Participant id: _____

Researcher observations	Participant quotes
Questions Comments Behaviors	

☐ **Sorting complete**

Categories

1. _____
2. _____
3. _____
4. _____
5. _____

Attributes of categories				
1	2	3	4	5

☐ **Camera data complete**

____ Has used at least one camera from the following:

____ point-and-shoot   ____ video camera

____ SLR   ____ uses phone as camera

____ web cam   ____ No camera or doesn't take picture

APPENDIX D: CAMERA STUDY SCRIPT

Arrival, getting started

[When the participant arrives, she'll sign in at reception.]

Researcher introduces the session

[Researcher greets participant, escorts participant to the testing room.]

[Say:] Okay, please come this way. Have a seat.

Thank you for agreeing to participate in this study. Your participation in the study is confidential. The data you provide will not have your name on it.

[Have Informed Consent form available.]

Before we get started, we have some paperwork to do. This is an Informed Consent Form for the first part of the session. [Hand to participant.] Please read it. It explains:

- What we are studying
- What you will do
- How the information will be treated
- And so on.

Please read both sides.

When you have finished reading it, if you are comfortable with what it says, please sign it. [Wait while participant reads and signs.]

Do you have any questions at this point?

I'll sign here now. My signature just says that I saw you read and sign the form.

Let's move over here now. [Researcher settles participant in front of the PC with the images on it.]

[Researcher guides participants to demographic questionnaire.]

Have a seat in front of the computer and I'll get us to a place where we can start. OK. For Subject Number, put in the number on your tag.

Demographic survey

Now, please fill in the questionnaire on the screen.

[Participant answers demographic questionnaire.]

(See Appendix A: Demographic Questions for contents of demographic questionnaire.)

Camera experiment

[Ideally, the participant will be settled at the PC, having filled in the demographic questionnaire.]

Researcher introduces camera affordances portion

[Researcher instructs participants to continue at the PC to identify images that are (and are not) cameras.]

[Say:]

Now we want to learn from you what makes something a camera. After you click Start, this PC will show you an image. You must answer the question at the bottom of the screen *as quickly as possible* to go to the next image. There will be 50 images. Answer the question for each image you are shown, again, as quickly as you can. When you're done, please tell me.

[Participant identifies cameras]

☐ Cameras identified

Researcher observations	Participant quotes

Questions

Comments

Behaviors

Image sorting

[Researcher explains sorting exercise.]

Thank you. Now I have another way for you to look at the images. Here is a stack of the same images you just looked at, on paper. Please sort the images into categories. You can have up to 5 categories, but you don't have to have that many. Ready?

You have 5 minutes. Go ahead.

[Researcher gives participant decks of images to sort.]

[Participant sorts deck into 3-5 categories.]

[Researcher instructs participant to write a name for each category.]

Now, please use a sticky and write a name for each category.

[Participant names each category.]

> [If participant does not sort based on camera attributes, say:] Thanks for doing that. Remember the last exercise where you answered whether the image you saw was a camera?
>
> When you looked at those images, how did you decide what was a camera and what wasn't?
>
> What would you say are the three main things that make something a camera?

[TIME: If the participant is over 3 minutes and it appears that she's going to take much longer, remind her at about 4:00 that she has 1:00 left. Say:]

You have one more minute.

[Researcher interviews participant about top 3 attributes of each category.]

Thank you for doing that. Please tell me about how you decided on the categories.

☐ Sorting complete

Categories

1. _____

2. _____
3. _____
4. _____
5. _____

Attributes of categories				
1	2	3	4	5

☐ **Camera data complete**

[If necessary, ask:] **What are the three main things that distinguish each category?**

[Researcher interviews participant about experiences related to cameras.]

Tell me about your experiences with cameras before today.

　　　____ Has used at least one camera from the following:

　　　____ point-and-shoot　　　　____ video camera

　　　____ SLR　　　　　　　　　　 ____ uses phone as camera

　　　____ web cam　　　　　　　　 ____ No camera or doesn't take picture

Thank you for that.

[Researcher interviews participant about how the participant approached the camera identification task.]

Thinking back to the task you completed on the computer with the 50 images, I am interested in knowing how you decided which images were of cameras and which were not cameras.

☐ **Experience data complete**

Great. Thank you for your participation.

[Researcher escorts the participant to reception.]

APPENDIX E: WHERE ARE VISITORS TO THE US COMING FROM?

DHS has many classifications of "nonimmigrant," including temporary visitors, aliens in transit, students, temporary workers, and exchange visitors. For purposes of this study, we focus on the non-resident visitors, which comprised 89.8% of the admissions in 2009, as reported by DHS [2]. These non-resident admissions were individuals who came to the US for a relatively brief period. Unknown (0.5%), e.g., not classified, short-term and long-term resident alien admission figures comprise 10.2% of the total admissions and are not discussed here.

In 2009, DHS recorded 32,544,098 admissions of temporary visitors, e.g., non-resident, nonimmigrant admissions.[4] 27,800,027 (or 85.4%) of those temporary visitors said they were visiting for pleasure, 4,390,888 (or 13.5%) said they were visiting for business, 346,695 (or 1.1%) were classified as transient aliens, and the remaining 6,488 (or 0.1%) were listed as commuter students.

Of all non-resident admissions, 60% were of people aged 25-54. People age 55 and older made up another 21% of admissions. Temporary visitors were almost evenly split between males and females.

Nearly half of all admissions for non-residents came primarily from the following countries:

Mexico	19.0%
United Kingdom	13.8%
Japan	9.6%
Germany	5.8%
Total	48.2%

The remaining admissions came from these countries:

---

[4] Note that "admissions" refers to events, not people: some individuals could have been admitted to the US multiple times during 2009.

France	4.8%
Italy	3.0%
Brazil	2.7%
Australia	2.4%
Netherlands	2.2%
Spain	2.2%
Other	33.9%
Unknown	0.6%
Total	51.8%

Note that roughly a third of admissions are from the broad "Other" category. We have narrowed that category down to all the countries of origin from which DHS reported more than 100,000 admissions each in 2009. These 23 countries, and the number of admissions from each, are listed below [5]:

India	574,568
China	512,386
Venezuela	480,623
Colombia	456,778
South Korea	410,904
Argentina	355,504
Israel	323,523
Bahamas	265,075
Dominican Republic	231,861
Jamaica	216,978
Taiwan	201,126
Philippines	200,884

Guatemala	192,281
Peru	181,164
Ecuador	163,440
Costa Rica	165,330
Trinidad and Tobago	159,633
Russia	146,985
Poland	133,591
El Salvador	133,453
Chile	130,367
Honduras	124,777
Panama	103,330

APPENDIX F: TECHNOLOGY ADOPTION BY COUNTRY

The World Bank and the International Telecommunications Union track and publish statistics on rates of technology adoption in countries around the world [1][6].[5] For our purposes, we are interested in three categories of technology adoption: personal computer ownership, Internet use, and mobile phone subscriptions.

---

[5] Statistics on Internet use and mobile phone subscriptions comes directly from the World Bank's World Development Indicators (WDI) database. Statistics on personal computer ownership come from the EconStats™ database and were aggregated from World Bank and International Telecommunications Union data.

Table 7 below displays the number of each of these items (per 100 people) in the US and the 15 countries from which the most temporary admissions to the US originate (see Appendix E: Where are Visitors to the US Coming From?). Using this data, we can draw some inferences about foreign visitors' average level of exposure to technology, and how their level of exposure compares to that of our study population.

Some countries on the list are comparable to the US in certain categories of technology adoption (e.g., personal computer ownership). In such cases, the relevant cell is colored blue. Where a country's level of technology adoption exceeds the US in a certain category, the relevant cell is colored green.

Table 7: Technology adoption statistics for the US and top 15 visiting countries

Country	2009 Temporary Admissions	Number of users per 100 residents		
		Personal Computers	Internet Users	Mobile Phone Subscribers
US		80.66	74.2	90.2
Mexico	6,168,774	14.4	31.1	80.6
United Kingdom	4,504,786	80.2	84.7	130.3
Japan	3,138,650	-	77.6	94.7
Germany	1,881,944	65.6	82.5	127.9
France	1,563,993	65.2	77.5	97.4
Italy	981,715	36.7	53.7	149.8
Brazil	869,310	16.1	40.7	104.1
Australia	775,885	-	75.9	100.9
Netherlands	715,023	91.2	90.7	115.4
Spain	713,310	39.3	65.8	112.0
India	574,568	3.3[7]	7.5	61.4
China	512,386	5.7	34.4	64.2
Venezuela	480,623	9.3[8]	35.9	96.7
Colombia	456,778	11.2[9]	36.5	96.1
South Korea	410,904	57.6	82.5	103.9

---

[6] 2006 data

[7] 2007 data

[8] 2005 data

[9] 2008 data

Only the Netherlands surpasses the US in personal computer, Internet, and phone use. Only the United Kingdom is on par with the US in terms of personal computer ownership, but it and a number of other countries – Japan, Germany, France, Australia, and South Korea – surpass the US in Internet users. Finally, most countries on the list above surpass the US in terms of mobile phone subscriptions, except for Mexico, India, and China.